Science Matters

GLACIERS

Christine Webster

WEIGL PUBLISHERS INC.

Published by Weigl Publishers Inc.
350 5th Avenue, Suite 3304, PMB 6G
New York, NY USA 10118-0069
Web site: www.weigl.com

Library of Congress Cataloging-in-Publication Data

Webster, Christine.
Glaciers / by Christine Webster.
p. cm. -- (Science matters)
Includes index.
ISBN 1-59036-303-5 (hard cover : alk. paper) -- ISBN 1-59036-309-4 (soft cover : alk. paper)
1. Glaciers--Juvenile literature. I. Title. II. Series.
GB2403.8.W45 2006
551.31'2--dc22

2004029145

Printed in the United States of America
1 2 3 4 5 6 7 8 9 0 09 08 07 06 05

Project Coordinator Tina Schwartzenberger
Copy Editor Janice L. Redlin
Design Terry Paulhus **Layout** Kathryn Livingstone
Photo Researcher Jason Novak
Consultant Roger Paulen, Alberta Geological Survey

Photograph Credits
Every reasonable effort has been made to trace ownership and to obtain permission to reprint copyright material. The publishers would be pleased to have any errors or omissions brought to their attention so that they may be corrected in subsequent printings.

Cover: Antartica Glacier from Angela Scott/Taxi/Getty Images
Jeff Brown: pages 7, 12-13; **Getty Images:** pages 1 (Brand X Pictures), 3T (Robert Harding World Imagery), 3M (Imagebank), 3B (Imagebank), 4 (Stone/John Turner), 6 (National Geographic/Gordon Wiltsie), 8 (Stone/Paul Wakefield), 10 (Taxi/Chris Salvo), 11 (Taxi/Angela Scott), 15 (The Image Bank/Michael Melford), 16 (Lonely Planet/Anders Blomqvist), 17 (NASA/AFP), 18 (The Image Bank/J.W. Burkey), 19 (Vin Morgan/AFP), 21 (Photodisc Green/Spike Mafford), 22T (Robert Harding World Imagery), 22B (Imagebank), 23T (Imagebank), 23B (Stone);
Roger Paulen, Alberta Geological Survey: page 14.

Contents

Studying Glaciers

A large mass of ice in a very cold region is called a glacier. Glaciers form in areas where it is so cold that snow does not melt. Layers of snow build up over many years and become ice. Glaciers can be as large as an entire **continent**. They can also fill a small valley between mountains. Over time, the glacier's weight and **gravity** cause it to move. Most glaciers move very slowly.

- If a glacier moves too quickly, the ice can crack.

Glacier Facts

Ice worms live on glaciers. These brown worms are 0.4 to 1.2 inches (1 to 3 centimeters) long. Keep reading to learn more about glaciers.

- Glaciers currently cover 5.8 million square miles (15 million square kilometers) of Earth.
- Eighty percent of Earth's fresh water is in glaciers.
- Some glaciers move more than 100 feet (30 meters) per day. Other glaciers only move about 6 inches (15 cm) per day.
- In 1827, a Swiss scientist placed a hut on a glacier. He returned 3 years later. The hut had moved 100 yards (91 m) downhill.
- Glaciers cover more than 46,603 square miles (75,000 square kilometers) of the United States. Most glaciers are located in Alaska.
- When glaciers melt, eskers sometimes form. Eskers are former rivers caused by melting water in glaciers that fill with sediment. When the glaciers melt away, large ridges are left behind.

How Glaciers Form

Glaciers only form in certain conditions. Snow falls in very cold, moist air. Snow is light and fluffy when it first falls. As more snow falls, the air in the bottom layer of snow disappears. The snow becomes ice. In very cold **climates**, thick layers of ice build up. Its weight and gravity cause the ice to move, or flow. The ice becomes a glacier.

■ Glacial ice covers about 10 percent of Earth's land area. Most of this glacial ice formed about 5,000 years ago.

The Glacier's Body

Most glaciers have three areas. One area is where the glacier grows larger. In the other area, glaciers become smaller.

Snow piles up in the upper part of a glacier. This area is called the accumulation zone.

Ablation is the loss of snow, ice, and water from a glacier. More material is lost than added in the lower part of a glacier. This area is called the ablation zone.

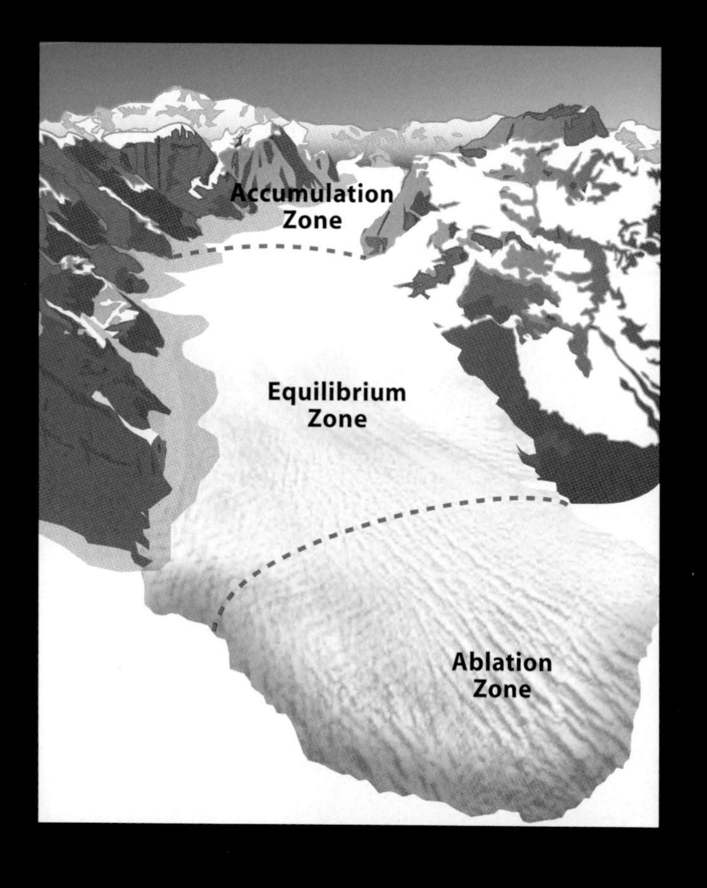

The area between the accumulation zone and the ablation zone is called the equilibrium zone. Very little snow, ice, or water is added or lost in this area.

Types of Glaciers

Glaciers can be grouped and sorted in different ways. One way to group them is based on their size. There are three sizes.

Ice Sheets

Ice sheets are the largest ice masses on Earth. They usually cover vast land areas. The Antarctic Ice Sheet spreads across 5 million square miles (13 million square km). It is more than 14,000 feet (4 km) thick.

Icecaps

Icecaps are smaller than ice sheets. When icecaps grow thick, ice overflows and travels down through glaciers in valleys. Icecaps are found in Alaska, the Canadian Arctic, Greenland, Iceland, and Norway.

Ice Fields

Ice fields form where mountain peaks and ridges push through glaciers in valleys. Ice fields are common in Alaska. One ice field can cover 1,500 square miles (4,000 square km).

- Mountain peaks that stick through ice are called *nunataks*.

Location, Location

Glaciers can also be grouped by their location.

Valley Glaciers

Valley glaciers form from icecaps high in mountains. They easily **erode** the landscape. Valley glaciers create U-shaped valleys. Some valley glaciers can be up to 3,000 feet (914 m) thick and 100 miles (161 km) long.

Alpine Glaciers

Alpine glaciers are sometimes called mountain or cirque glaciers. A cirque is a rounded, bowl-shaped area where snow collects. Alpine glaciers are located high in the mountains. They are smaller than valley glaciers. Ice easily flows over the cirque and travels down mountains into valleys. Alpine glaciers can cause **avalanches**.

Piedmont Glaciers

Piedmont glaciers are located at the bottom of mountain ranges. Piedmont glaciers are wide and round. They form when valley glaciers flow over lower mountain slopes and spread out.

Glacial Movement

Glaciers move in two different ways. They slide and creep. Sometimes ice layers slide past each other. For instance, if a bottom layer melts, a thin layer of water forms where the ice was. This allows the glacier to slide more easily. This is called sliding.

The other form of movement is called creep. The glacier is so heavy that ice layers form on top of other ice layers. The weight of the top layers changes the glacier's shape inside. The weight also causes the glacier to move.

■ Glaciers move faster or slower depending on the season, how much snow falls, and the glacier's mass.

Crevasses

As glaciers move, they stretch and twist. Movement causes the ice to crack. The cracks are called crevasses. Crevasses can be more than 100 feet (30 m) deep and more than 60 feet (18 m) wide.

Crevasses are dangerous. Light snow can cover and hide them. People and animals can easily fall into a crevasse. Scientists use crevasses to study Earth. Glacier ice inside crevasses tells scientists what Earth was like hundreds and thousands of years ago.

■ Crevasses are the greatest danger to people working on glaciers.

The Water Cycle

Earth has a limited amount of water. Water **recycles** itself through the water cycle. Glaciers are part of the water cycle. Water from melting glaciers flows into lakes, rivers, and oceans. This diagram shows how the water cycle works.

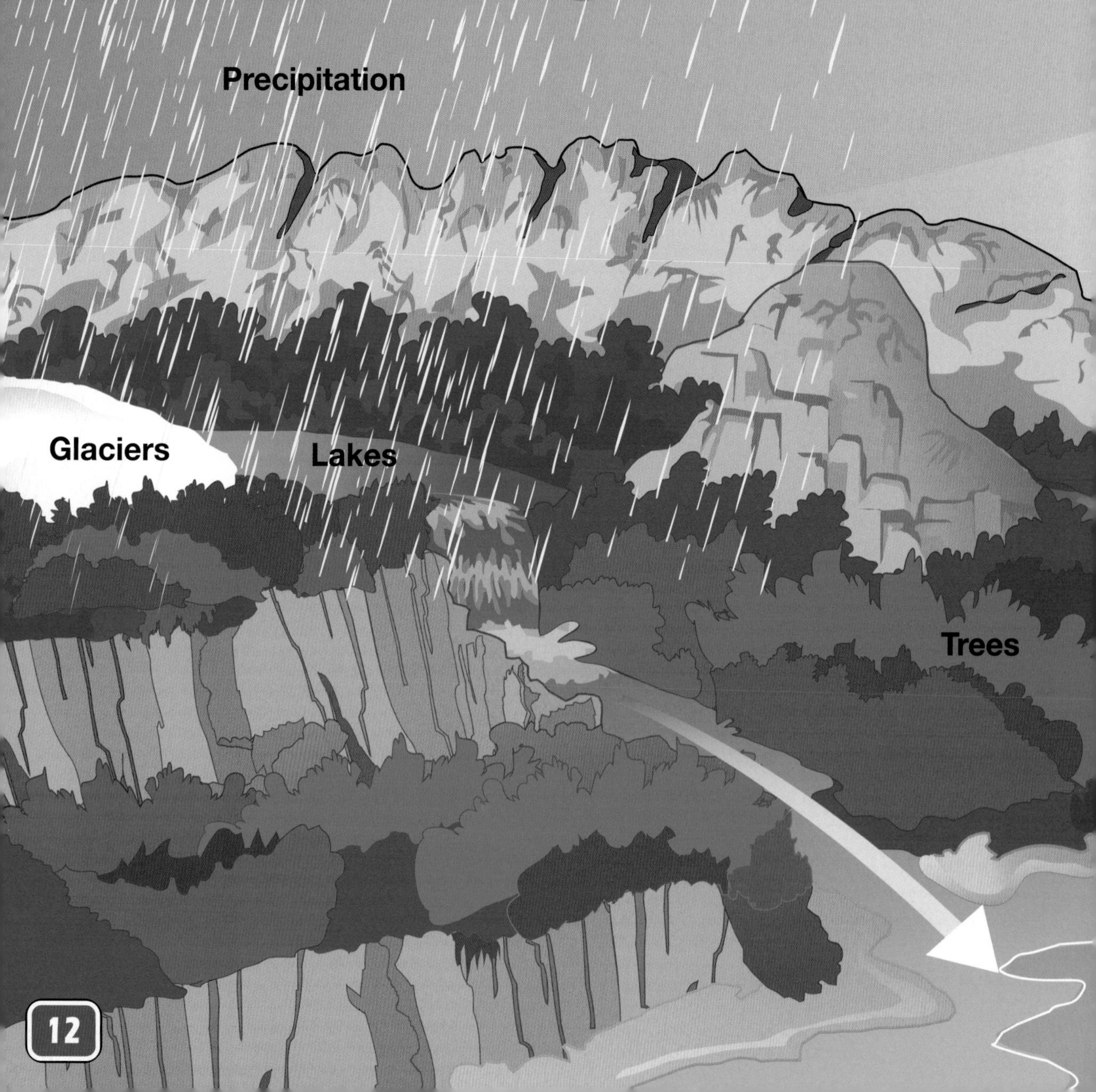

Precipitation

Glaciers

Lakes

Trees

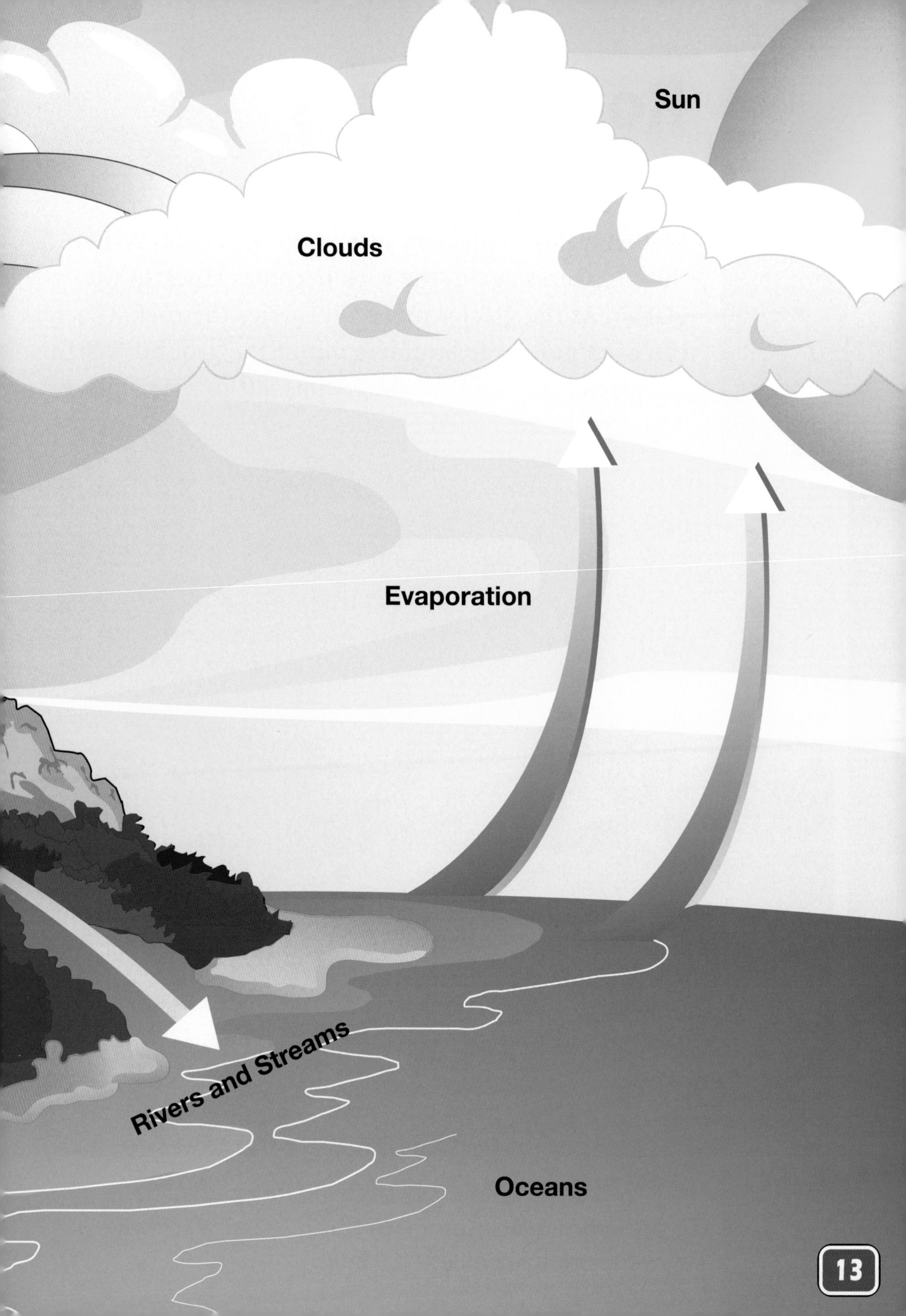
Sun
Clouds
Evaporation
Rivers and Streams
Oceans

Rock Movers

Glaciers are powerful. Most rocks have small cracks. Melted ice from glaciers seeps into the cracks. When the ice freezes again, the rock becomes stuck to the glacier. As the glacier moves, it carries the rock. Glaciers can even pull large boulders out of the ground. When the glacier melts again, the rock is left in a new place. These rocks are called erratics.

■ The Okotoks Erratic is located in Alberta, Canada. It is the largest glacial erratic in North America. It likely moved about 248 miles (400 km).

Gouging

Rocks stuck in glaciers can gouge, or cut, the land with deep grooves as the glacier travels. Gouging changes the landscape.

Sometimes glaciers gouge valleys near coastlines. Some valley glaciers reach coasts. As the ice melts, the glacier moves back. Sometimes the glacier leaves behind a *fjord*. A fjord is a narrow inlet with steep mountains on either side. Glaciers can also gouge the sides of mountains. After many years, the mountain top can resemble a horn or sharp peak.

The Twin Sawyer glaciers formed the Tracy Arm Fjord in Alaska.

Glacial Deposits

Glaciers can carry thousands of small rocks as they travel. On the journey, the rocks break down into smaller pieces. This small, fine-grained material is called rock flour. Rock flour mixes with **sediment** and water. As a glacier melts, it deposits debris, called *till*. A mixture of rock flour, rocks, and other material can be picked up and carried by the glacier. Glacial deposits build up into ridges or piles called moraines.

■ There are eight types of moraine. Only two types exist before the glacier melts.

Glacial Lakes

Glaciers can create lakes. Many glacial lakes have formed after glaciers melt into holes in the landscape. The five Great Lakes are former glacial lakes.

One specific type of glacial lake is called a kettle. Kettles form when blocks of ice are buried by glacial debris. The blocks of ice melt very slowly, and a depression, or low area, forms. The depression fills with water, forming a lake. Kettles are usually very small.

■ About 18,000 years ago, a glacier covered almost all of Canada and some of the United States. This glacier formed the Great Lakes.

Icebergs

An iceberg is a piece of a glacier that has broken off and floated out to sea. The process of an iceberg breaking off a glacier is called calving. Icebergs can be many miles long. Over time, icebergs melt in warm ocean waters.

Icebergs are dangerous. More than 75 percent of an iceberg is underwater. In 1912, the ship, *Titanic*, struck an iceberg. About 1,513 of the 2,220 passengers on the ship died. Today, coast guards patrol the oceans for icebergs to prevent this from happening again.

■ The largest icebergs form around Antarctica. The tallest icebergs form in the northern part of the Atlantic Ocean.

Glaciologist

A glaciologist is someone who studies glaciers. By studying a glacier's growth and shrinkage, a glaciologist learns about Earth's climate.

Glaciologists work in freezing conditions. They often sleep in tents in the snow. Sometimes they work around crevasses. For safety, glaciologists wear harnesses in case they fall into a crevasse. Glaciologists use different tools to measure how glaciers change. They take samples of glacial ice called ice cores. Ice cores show the annual growth of ice layers. Ice cores show when Earth's temperature was warmer or cooler.

Surfing Water Science

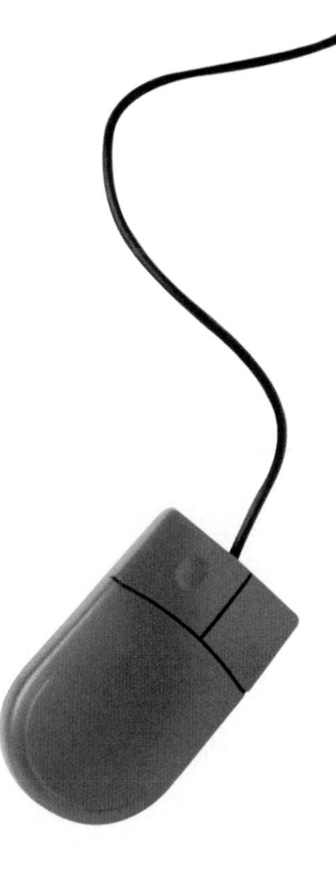

How can I find more information about glaciers?

- Libraries have many interesting books about glaciers.
- Science centers and museums are great places to learn about glaciers.
- The Internet offers some great Web sites dedicated to glaciers.

Where can I find a good reference Web site to learn more about glaciers?

Encarta Homepage
www.encarta.com

- Type any glacier-related term into the search engine. Some terms to try include "crevasse" and "iceberg."

How can I find out more about glaciers and icebergs?

All About Glaciers
http://nsidc.org/glaciers

- This Web site offers a detailed look into the life of a glacier.

Science in Action

Rising Sea Levels

Try this experiment to see how melting glaciers can affect sea levels.

Tools Needed:

- 2 plastic bowls
- water
- ice
- ruler
- piece of paper and pencil

Directions:

1. Fill one plastic bowl with water. Place the bowl in the freezer.
2. When the water has completely frozen, pop the ice out of the plastic bowl.
3. Place the frozen block of ice in the second plastic bowl.
4. Slowly add water to the ice until the ice begins to float.
5. With your ruler, measure the level of the water in the bowl. Record your measurement on a piece of paper.
6. Allow the ice to melt completely.
7. Measure the water depth again. Compare your results to the first measurement.

Did the "sea" level rise? How do you think melting glaciers can affect the level of water in the oceans?

What Have You Learned?

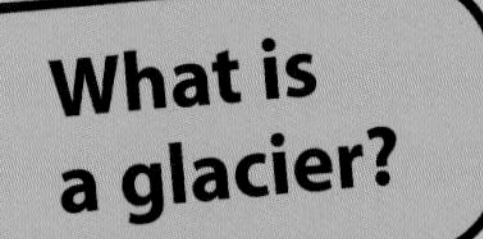

1. What is a glacier?

2. What causes glaciers to move?

3. What was the name of the ship that sank after hitting an iceberg in 1912?

4. In how many ways can glaciers be grouped?

5. Are ice sheets and icecaps the same?

6 What is a rock that is carried by a glacier called?

7 What type of valleys do glaciers usually gouge?

8 What is a piece of a glacier that has broken off and floated out to sea called?

9 What is a glaciologist?

10 What can glaciers tell us about our world?

Answers: 1. A glacier is a large mass of ice in a very cold region. **2.** Weight and gravity **3.** *Titanic* **4.** Two. Glaciers can be grouped by location and by size. **5.** No. Ice sheets are bigger than icecaps. **6.** An erratic **7.** Glaciers usually gouge U-shaped valleys. **8.** An iceberg **9.** Someone who studies ice and snow **10.** Glaciers tell us whether our climate is changing.

Words to Know

avalanches: sudden sliding of large masses of snow and rock down a mountain

climates: the usual weather in a region throughout the year

continent: the seven large land masses on Earth—Africa, Antarctica, Asia, Australia, Europe, North America, and South America

erode: remove rock and pieces of soil by natural forces such as water, ice, waves, and wind

gravity: the force that pulls objects toward the center of Earth

recycles: returns to an original condition so a process can begin again

sediment: very small pieces of rock and dirt deposited by water, wind, or ice

Index